INTERNATIONAL BANK FOR RECONSTRUCTION AND DEVELOPMENT

WORLD BANK

WORLD BANK STAFF OCCASIONAL PAPERS NUMBER THIRTEEN

HERMAN G. VAN DER TAK
and
ANANDARUP RAY

THE ECONOMIC BENEFITS OF ROAD TRANSPORT PROJECTS

Distributed by The Johns Hopkins Press
Baltimore and London

Manufactured in the United States of America
Library of Congress Catalog Card Number 79-152911

ISBN 0-8018-1314-x

FOREWORD

I would like to explain *why* the World Bank Group does research work, and why it publishes it. We feel an obligation to look beyond the projects we help to finance toward the whole resource allocation of an economy, and the effectiveness of the use of those resources. Our major concern, in dealings with member countries, is that all scarce resources, including capital, skilled labor, enterprise and know-how, should be used to their best advantage. We want to see policies that encourage appropriate increases in the supply of savings, whether domestic or international. Finally, we are required by our Articles, as well as by inclination, to use objective economic criteria in all our judgments.

These are our preoccupations, and these, one way or another, are the subjects of most of our research work. Clearly, they are also the proper concern of anyone who is interested in promoting development, and so we seek to make our research papers widely available. In doing so, we have to take the risk of being misunderstood. Although these studies are published by the Bank, the views expressed and the methods explored should not necessarily be considered to represent the Bank's views or policies. Rather they are offered as a modest contribution to the great discussion on how to advance the economic development of the underdeveloped world.

ROBERT S. MCNAMARA
President
International Bank for
Reconstruction and Development

TABLE OF CONTENTS

FIGURES

PREFACE

The World Bank takes a considerable interest in developing the transport sector of the less developed countries' economies. Appraising transport projects raises many different kinds of problems, and the World Bank Staff Occasional papers have dealt with several of these ranging from methods of quantification to an account of the economics of road user charges.

The present paper explores, for the benefit of those who have to appraise road projects, the comprehensiveness and applicability of one widely used method for estimating benefits, namely the measure of social surplus. Some simple versions of the social surplus method are discussed in detail to clarify the nature of the benefits covered by such measures. The method is also applied to more complex situations in an effort to indicate the range as well as the limitations of its usefulness, and to help show how it can be adapted in practice. The presentation uses only elementary algebra and geometry.

The paper's principal use will be to help those, both within the Bank and outside, who have to appraise road projects, to apply the method in a variety of new situations.

It is impracticable to thank individually all those who have contributed to this piece of work, but special thanks are due to Mr. Benjamin B. King for most helpful criticism, to Professor A. A. Walters, and also to the authors' colleagues in the Sector and Projects Studies Division. The views expressed in this paper are those of the authors, however, and they alone are responsible for them.

A. M. Kamarck
Director
Economics Department

THE ECONOMIC BENEFITS OF ROAD TRANSPORT PROJECTS

I

INTRODUCTION

The economic evaluation of a project in any sector entails the measurement and comparison of cost and benefit streams expected from alternative investments. This paper presents an exposition of the social surplus method of measuring benefits. In this method, used within as well as outside the Bank, benefits are measured in terms of the concepts of consumers' and producers' surpluses. Though the exposition is developed in terms of road investments, the method's use and the paper's relevance are not restricted to road transport projects.

The exposition is intended to shed light on the nature of benefits to be expected from road transport projects, both with and without various types of market imperfections, and in particular, to show how these benefits relate to changes in the supply and demand of transported commodities. The method we present is designed for analyzing road projects in isolation from other investments. If a road project is a component of an investment package, it may not be possible to assign benefits meaningfully to road alone. Benefits from alternative packages, including road projects, should be conjointly measured in terms of the whole program's ultimate objectives—such as the development of farm output, tourism, etc.

Throughout, the paper is concerned with basic concepts which frequently give rise to confusion in applied economic work in this area. Heavily stylized examples using linear supply and demand functions are employed to explain these concepts, proceeding from simple illustrations to more complex and more realistic cases. We recognize that the interrelations of various roads, modes and regions are more complex than here portrayed, and that market imperfections are, in practice, more often the rule than the exception. We feel, however, that the stylized examples capture the essence of various important relationships bearing on the benefits from road improvements, and that analysis of these cases contributes to a better understanding of the issues involved. We hope to provide a better basis for the judgments, adjustments and measurements that need to be made in project appraisal.

Thus one purpose of the paper is purely expository: to record the economic mechanics behind the social surplus method of benefit evaluation as applied to road projects. We also hope, however, that even for those familiar with social surplus, the paper will indicate how the method can be adapted to cases for which the conventional application is not realistic, and will provide some useful insight into benefit valuation.

The paper provides an analytical framework and touches only briefly on the problems of filling the framework with figures. It is not concerned with the empirical specification of cost and demand functions which must precede the measurement of benefits,[1] nor does it discuss the criteria for judging the desirability of projects; it deals only with the benefit measurement stage of the project evaluation process.[2]

In some cases the benefits from a road transport project in any year can simply be measured by the product of the project-induced decrease in unit road user costs and the normal volume of traffic.[3] This measure will be valid only when the volume of traffic on the improved road is not responsive to changes in the unit transport cost. In general, however, traffic volume will increase with road improvement (as when a new road opens up an isolated region) and measuring benefits only in terms of normal traffic will underestimate the benefits. In the following chapters we will present a method of measuring benefits which takes traffic responses into account, discusses the factors underlying their relative importance, relates

[1] The issue of measuring road user costs is addressed in World Bank Staff Occasional Paper No. 2, Jan de Weille's *Quantification of Road User Savings* (Baltimore: Johns Hopkins Press, 1966).

[2] For a general description of the project evaluation process in transportation, see Hans A. Adler, *Sector and Project Planning in Transportation*, World Bank Staff Occasional Paper No. 4 (Baltimore: Johns Hopkins Press, 1967), Chapter II, "Project Planning."

[3] The use of "normal" to designate project-independent traffic flows is standard. See, for example, Adler, *Sector and Project Planning*, p. 45.

the benefits to associated changes in the production and consumption of the transported commodity, and interprets the impact of market imperfections on benefit measurement.

The exposition of the social surplus method is developed in stages, beginning with very simple regional models which assume the absence of competing transport modes, market imperfections and externalities. In Chapter II, a two-regional model is first discussed. A third region is then introduced to indicate that the model's use is not affected by a multiplicity of regions. Finally, the effects of an increasing transport cost function are considered. In Chapter III competing roads and transport modes are introduced into the model. Chapter IV discusses how market imperfections and other externalities would change the results of the analysis, and Chapter V summarizes the paper's conclusions.

II

MEASURING BENEFITS IN THE ABSENCE OF COMPETING TRANSPORT

In this chapter we examine, with the help of simple models, the relationship of the benefits from a road improvement project to changes in the pattern and volume of regional production and consumption when the improved road is the only transport route connecting the regions. While this simplification may seem exaggerated, it permits us to consider benefits to normal traffic (traffic without the project) and induced traffic (new traffic) without having to be concerned about traffic diverted to the improved road from competing routes. The next chapter deals with the more general case where competing routes are available.

In the first section we develop a model with two trading regions and a constant transport cost function. This enables us to formulate the social surplus measure in the conventional way. A third region is introduced in the second section to indicate the generality of the method.[1] In the third section we consider the effect of allowing the unit supply price of road transport to increase with an increase in the volume of traffic. Throughout the paper it is assumed that producers' and consumers' gains and losses can

[1] The model described in these two sections follows closely that found in a well-known article by P. Samuelson, "Spatial Price Equilibrium and Linear Programming," *American Economic Review* (June, 1952), in which he formulates a linear program for the solution of the multi-regional trade allocation problem.

be added and subtracted,[2] a standard restriction for the application of the social surplus method.

A Simple Two-Region Model

In this model two regions are connected by a single road, used exclusively to transport one commodity traded by the two regions. The importing region is *A* and the exporting region, *B*. Without the planned improvement, the excess of the price in *A* over that in *B* equals the preexisting unit transport cost, which is assumed invariant with respect to the volume of transport measured in units of the commodity. Similarly, the price differential in the equilibrium after the improvement equals the new and lower unit transport cost, which is again assumed to be a constant. The change in the unit transport cost will thus equal the change in the price in region *A* minus the change in price in *B*. If θ is unit transport cost and P is the price of the commodity, with subscripts denoting regions, then:[3]

$$\Delta\theta = \Delta P_A - \Delta P_B$$

The relationship between the unit transport cost on the road and the volume of transport between *A* and *B* will be referred to as the demand function for transport between *A* and *B*. This demand function will have a negative slope, derived at any given unit transport cost from the slopes at the corresponding prices of the demand and supply curves for the com-

[2] The producers' and consumers' gains and losses can be weighted to take account of personal and regional income distribution effects, as long as the weights are predetermined by the policy-makers.

[3] All changes, expressed by the Greek letter delta, are defined as new value minus old value. Thus, if $P_A{}^0$ and $P_A{}'$ are the price levels in *A* in the old and new equilibria, respectively, and the corresponding price levels in *B* are $P_B{}^0$ and $P_B{}'$ then

$$\Delta P_A = P_A{}' - P_A{}^0$$

and

$$\Delta P_B = P_B{}' - P_B{}^0.$$

If θ^0 and θ' are the levels of the unit transport cost in the two situations, then the equilibrium conditions are

$$\theta^0 = P_A{}^0 - P_B{}^0$$

and

$$\theta' = P_A{}' - P_B{}'.$$

Hence,

$$\begin{aligned}\Delta\theta &= \theta' - \theta^0\\ &= \Delta P_A - \Delta P_B.\end{aligned}$$

FIGURE 1
THE TRANSPORT DEMAND FUNCTION

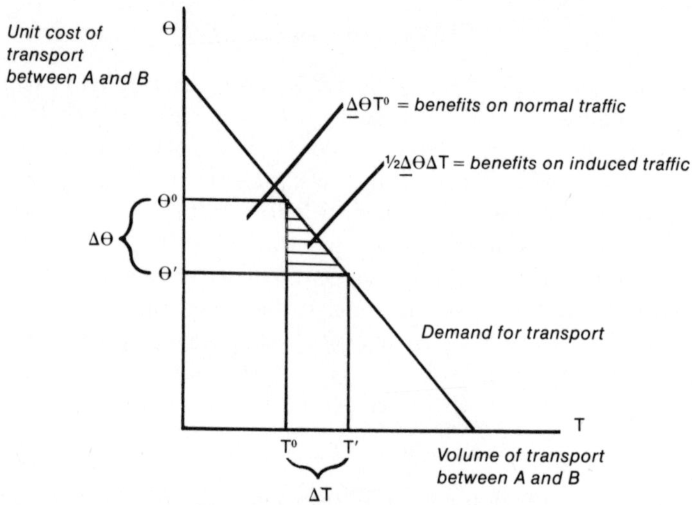

modity in the two regions. This relationship can also be expressed in terms of the corresponding price elasticities.

The change in the area under this transport demand curve (above the equilibrium price) due to the fall in the unit transport cost may here be regarded as a measure of the total benefits to the economy attributable to the project in the period concerned. The model demonstrates that this area represents the residual gains to the consumers and producers of the commodity in the two regions, that is, their gains net of losses.

The transport demand curve is shown in Figure 1. T^0 and T' are the equilibrium volumes of traffic without and with the improvement, respectively; $T' - T^0$ is denoted by ΔT and, as before, $\theta' - \theta^0$ is denoted by $\Delta\theta$. The change in the area under the demand curve (and above the equilibrium price) which is due to $\Delta\theta$ is $\underline{\Delta}\theta(T^0 + \frac{1}{2}\Delta T)$.[4] The term $\underline{\Delta}\theta T^0$ measures the

[4] Note that since $\Delta\theta$ is negative and the quantity $(T^0 + \frac{1}{2}\Delta T)$ is positive, their product will also be negative. It will represent the decrease in the area under the demand curve *below* the equilibrium price. However, benefits are measured by the *increase* in the area under the demand curve above the equilibrium price. This is a positive quantity and equals $\underline{\Delta}\theta(T^0 + \frac{1}{2}\Delta T)$ where $\underline{\Delta}\theta$ is the absolute (i.e. the positive) value of $\Delta\theta$ (this might also be indicated by the more usual notation of $|\Delta\theta|$). For example, if transport cost falls by five monetary units, then $\Delta\theta = -5.00$ but $\underline{\Delta}\theta = 5.00$. Whenever necessary, we shall indicate the absolute value of a negative change by the notation $\underline{\Delta}$ (delta bar).

FIGURE 2
THE MARKET FOR A TRANSPORTED COMMODITY IN REGIONS A AND B

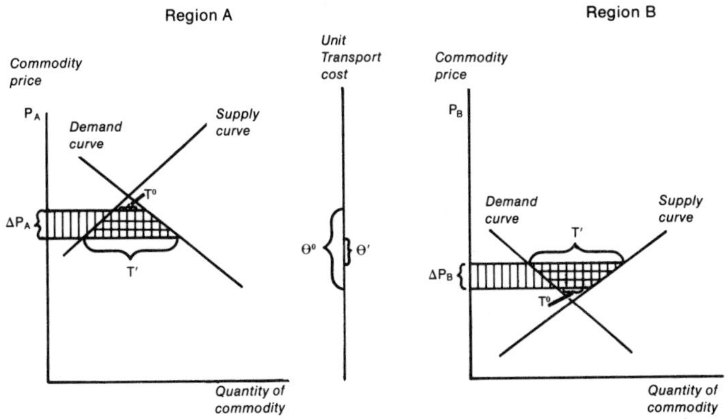

positive value of the fall in the total transport cost, or cost savings, on the normal volume of traffic, T^0.[5] The term $\frac{1}{2}\Delta\theta\Delta T$ is the shaded area and represents, as will be seen below, the benefits from the induced traffic, ΔT, that is, new traffic attracted to the road by the improvement.

Figure 2 shows the corresponding supply and demand equilibria for the commodity in the two regions without and with the road improvement. The price in A falls and the price in B rises when the unit transport cost falls. At the same time, the volume of traffic in the traded commodity increases from T^0 to T'. The increase in traffic equals the excess demand in A and the excess supply in B.

To see how the slope of the transport demand function in Figure 1 is related to the slopes of the commodity demand and supply functions in the two regions (Figure 2), let us suppose that $-u$ and $-w$ are the negative slopes of the demand functions in regions A and B, respectively, and that v and x are the positive slopes of the supply functions in regions A and B, respectively.[6] In region A,

$$\frac{\Delta T}{\Delta P_A} = \frac{\Delta D_A}{\Delta P_A} - \frac{\Delta S_A}{\Delta P_A}$$
$$= -u - v$$

[5] It should be clear that *normal traffic* refers to the volume of traffic that would obtain *without* the project, not to the volume of traffic at the time the road is improved. In most cases, one may expect a growth in traffic even if the road is not improved, in which case the normal volume of traffic will be higher in later years.

[6] Following the convention in mathematics, we define the slope in this paper as the change in the dependent variable, i.e. quantity demanded or supplied, per unit change in the independent variable, i.e. price.

in other words,

$$\Delta T = -(u + v)\Delta P_A$$
$$= -L\Delta P_A$$

where

$$L = u + v$$

Similarly, in region B

$$\frac{\Delta T}{\Delta P_B} = \frac{\Delta S_B}{\Delta P_B} - \frac{\Delta D_B}{\Delta P_B}$$
$$= x - (-w)$$
$$= x + w$$

in other words,

$$\Delta T = (x + w)\Delta P_B$$
$$= M\Delta P_B$$

where

$$M = x + w$$

Since, however, $\Delta\theta$ equals ΔP_A minus ΔP_B, we have

$$\Delta\theta = -\frac{\Delta T}{L} - \frac{\Delta T}{M}$$
$$= -\Delta T \frac{L + M}{LM}$$

Thus, the slope of the transport demand function is given by

$$\frac{\Delta T}{\Delta\theta} = -\frac{LM}{L + M}$$

This expression for the slope of the transport demand function implies that, given the commodity demand functions in the two regions, the traffic generated by a unit fall in the transport cost will be greater, the greater the price responsiveness of either one or both commodity supply functions. Similarly, the amount of generated traffic will be greater, the greater the price responsiveness of either one or both commodity demand curves, given the commodity supply functions. If the initial volumes of production and consumption are known, one can translate this observation into terms of price elasticities. Again, *ceteris paribus*, the higher the absolute (or positive) value of the price elasticity of any of the four commodity demand

and supply curves at the initial point, the higher will be the absolute value of the elasticity of the transport demand curve at that point.[7]

The total benefits attributable to the project can be seen with reference to Figure 2. In region A, the horizontally shaded area represents the total gain in consumers' surplus due to the reduction in the price of the commodity in A. The vertically shaded area represents the loss in terms of producers' surplus. This loss offsets a part of the total consumers' surplus, leaving an area measured by $\underline{\Delta}P_A(T^0 + \frac{1}{2}\Delta T)$ as the residual gain to region A. Similarly, in region B, a part of the producers' surplus realized due to the price rise in B (vertically shaded area) is offset by the loss in terms of consumers' surplus (horizontally shaded area). The residual gain in B is measured by $\Delta P_B(T^0 + \frac{1}{2}\Delta T)$. The total residual gain is thus $(\underline{\Delta}P_A + \Delta P_B)(T^0 + \frac{1}{2}\Delta T)$, or $\underline{\Delta}\theta(T^0 + \frac{1}{2}\Delta T)$. This is the change in the area under the transport demand curve, as we showed above. We have thus established the equivalence between the measure of benefits defined by the change in the area under the transport demand curve and the measure of benefits defined in terms of consumers' and producers' surpluses in the two regions.[8]

Clearly this conclusion follows only if we are indifferent with respect to benefit distribution between the two regions and between producers and consumers (as specified in footnote 2 in this chapter). The change in the area under the demand curve for transport is *not* a measure of the total residual gains if different weights are attached to the gains and losses of producers and consumers, interregionally or interpersonally, in order to account for changes in income distribution. Such weighting renders the derived transport demand curve useless as a tool for benefit measurement.[9]

It is often convenient to regard the benefits $\underline{\Delta}\theta(T^0 + \frac{1}{2}\Delta T)$ as the sum

---

[7] If $E(T)$ is the elasticity of the transport demand curve, $E(D_A)$ and $E(S_A)$ are the elasticities of demand and supply for the commodity in A, $D_A{}^0$ and $S_A{}^0$ are the initial demand and supply volumes in A, and we define

$$K_A = -D_A{}^0\, E(D_A) + S_A{}^0\, E(S_A), \text{ and } K_B \text{ analogously, then}$$

$$E(T) = \frac{\theta^0}{T^0} \cdot \frac{\Delta T}{\Delta\theta} = -\frac{\theta^0}{T^0} \cdot \frac{K_A K_B}{P_B{}^0\, K_A + P_A{}^0\, K_B}$$

[8] If nonlinear demand and supply curves are used, the benefit measure loses its algebraic simplicity but the substance of the analysis is the same.

[9] This is especially important for transport projects affecting international traffic. With such projects the part of the benefits accruing to foreign countries, if any, may be irrelevant from the point of view of the country undertaking the project.

of two parts: $\underline{\Delta}\theta T^0$ and $\frac{1}{2}\underline{\Delta}\theta\Delta T$. The first part is the sum of the cost savings realized by the consumers in *A* and the increase in profits realized by producers in *B*, on the normal volume of traffic. The second part, which corresponds to the shaded area under the transport demand curve in Figure 1, measures the sum of the consumers' surplus in *A* and producers' surplus in *B* on induced traffic. This may be referred to as *induced benefits.*

The ratio of induced benefits to the benefits realized on the normal volume of traffic is $\frac{1}{2}\Delta\theta \cdot \Delta T/\Delta\theta \cdot T^0$ or $\frac{1}{2}\Delta T/T^0$, or $-\frac{1}{2}E(T)[\underline{\Delta}\theta/\theta^0]$, where $E(T)$ is the elasticity of the transport demand curve. Induced benefits will be larger in relation to the benefits on the normal volume of traffic (i.e. in relation to road user savings) the larger is either $E(T)$, or ΔT relative to T^0, or $\underline{\Delta}\theta$ relative to θ^0.

Intuitively, the reason for these results is quite clear. The volume of transport in this simple model is directly derived from the volume of trade. Apart from the scalar transformation of introducing a constant mileage, the demand curve for transport is simply the locus of the different equilibrium volumes of trade. The latter are reflections of the demand and supply conditions in the regions. The slope and elasticity of the transport demand curve depend therefore on the specific commodity demand and supply functions in the two regions. An improvement in transport between the two regions leads to lower transport costs, changes in production and consumption of the commodity transported, and more trade and transport between the two regions.

The road improvement may be measured in this case either as the sum of the changes in consumers' and producers' surplus on the commodity in the two regions, or as the increase in consumers' surplus under the transport demand curve. If income distribution effects are ignored, these two measures are equivalent: they are different ways of measuring the same quantity and should not be computed separately and added, as is sometimes done. The benefits consist of road user savings on the normal traffic plus benefits on induced traffic. The relative importance of induced benefits to road user savings is shown to be greater, the larger is the percentage reduction in transport costs and the larger the elasticity of demand for transport.

Multiple Regions

These results remain substantially unchanged regardless of the number of regions introduced in the model. Only the slope of the transport demand curve between regions *A* and *B* is altered. To see this, let us introduce a third region, *C*, which competes with *B* in supplying the commodity to *A*. When the road between *A* and *B* is improved, the competitive position of C worsens: transport increases between A and B and decreases between A and C. If the unit transport cost between A and C remains constant, then the price in A and C will fall by the same amount (i.e. $\Delta P_A = \Delta P_C$). The price in B rises. For simplicity, we assume that C remains an exporter to A throughout.

In this case, the following three conditions must hold for the equilibria without and with the improvement in the three regions:

(i)
$$\begin{aligned}\Delta\theta_{AB} &= \theta_{AB}' - \theta_{AB}^0 \\ &= \Delta P_A - \Delta P_B\end{aligned}$$

where θ_{AB} is the unit transport cost on the road between A and B;

(ii)
$$\Delta P_A = \Delta P_C$$

since the unit transport costs between A and C remains constant, and

(iii)
$$\Delta T = \Delta T_{AB} - \underline{\Delta} T_{AC}$$

that is, the change in the volume of traffic bringing the commodity into A equals the difference between the change in the traffic coming from B and that coming from C, because the increase in transport between A and B when the road connecting them is improved is partly at the expense of transport between A and C.

From these conditions one can again derive the slope of the transport demand curve between A and B:[10]

$$\frac{\Delta T_{AB}}{\Delta\theta_{AB}} = -\frac{M(L+N)}{L+M+N}$$

The same conclusions as in the two-region case regarding the relationships between the slopes and elasticities of the demand and supply curves in the

---

[10] Region A: $\begin{aligned}\Delta T &= \Delta T_{AB} - \underline{\Delta} T_{AC} \\ &= \Delta T_{AB} + \Delta T_{AC} \\ &= -L\Delta P_A\end{aligned}$

Region B: $\Delta T_{AB} = M\Delta P_B$

Region C: $\Delta T_{AC} = N\Delta P_C = N\Delta P_A$

where $N = z + y$, z and $-y$ being the slopes in C of the supply and demand for the commodity transported, respectively.

Therefore: $\begin{aligned}\Delta\theta_{AB} &= \Delta P_A - \Delta P_B \\ &= -\frac{\Delta T_{AB}}{L+N} - \frac{\Delta T_{AB}}{M} \\ &= -\Delta T_{AB}\frac{L+M+N}{M(L+N)}\end{aligned}$

Hence, the expression in the text.

regions, and between the slope and elasticity of the transport demand curve between A and B, hold in this case also.[11]

What about the benefits? There are now residual gains in A and B, as before, but a residual loss in C. The residual gain in A, accruing wholly to consumers, is in this case $\Delta P_A(T_{AB}{}^0 + \frac{1}{2}\Delta T_{AB}) + \Delta P_A(T_{AC}{}^0 - \frac{1}{2}\Delta T_{AC})$. Note that A gains on its normal transport from B and from C; and that A gains on induced traffic from B but loses on reduced traffic from C. In B producers enjoy a residual gain of $\Delta P_B(T_{AB}{}^0 + \frac{1}{2}\Delta T_{AB})$, while in C they suffer a residual loss of $\Delta P_C(T_{AC}{}^0 - \frac{1}{2}\Delta T_{AC})$. Since ΔP_A equals ΔP_C, this residual loss in C is offset by a part of the residual gain in A. Adding the remaining part of the gain in A to the residual gain in B, we have $\Delta P_A(T_{AB}{}^0 + \frac{1}{2}\Delta T_{AB}) + \Delta P_B(T_{AB}{}^0 + \frac{1}{2}\Delta T_{AB})$ as the total benefits due to the road project in this case. This, once again, equals the change in the area under the demand curve for transport between A and B,

$$\Delta\theta_{AB}(T_{AB}{}^0 + \tfrac{1}{2}\Delta T_{AB})$$

Thus the same equivalence continues to hold between the measurement of benefits in terms of the change in the area under the demand curve for transport (between A and B) and that in terms of changes in consumers' and producers' surpluses in the region concerned. The impact of the third region C is fully represented in the change of the slope of the transport demand curve between A and B, and there is no need for additional adjustments to allow for the contraction of output in C, or for the reduction in the volume of transport between C and A.

Consider again the composition and distribution of the benefits. Road user savings benefit consumers in A ($\Delta P_A\ T_{AB}{}^0$), and producers in B ($\Delta P_B\ T_{AB}{}^0$). Induced benefits accrue to consumers in A ($\frac{1}{2}\Delta P_A\ \Delta T_{AB}$), and producers in B ($\frac{1}{2}\Delta P_B\ \Delta T_{AB}$). This is the same story as in the two-region case. In addition, consumers in A now enjoy road user savings of $\Delta P_A\ T_{AC}{}^0$ minus the induced loss of $\frac{1}{2}\Delta P_A\ \Delta T_{AC}$, as a transfer payment from the producers in C. Furthermore, in each region there are transfers between consumers and producers, resulting in residual gains or losses in consumers' or producers' surpluses. Clearly, the equivalence of benefit measures is again valid only if we are not concerned with regional and personal income distribution.

[11] In this case,

$$E(T) = -\frac{\theta_{AB}{}^0}{T_{AB}{}^0} \cdot \frac{K_A\,K_B\,P_C{}^0 + K_B\,K_C\,P_A{}^0}{P_C{}^0(K_A\,P_B{}^0 + K_B\,P_A{}^0) + K_C\,P_A{}^0\,P_B{}^0}$$

where K_A and K_B are as in footnote 7 to this chapter, and K_C is defined correspondingly.

These general results are not dependent on the fact that C is assumed to be an exporter. The same equivalence between the two benefit measures obtains if C is assumed to be a competing importer. The size and distribution of the benefits will, of course, differ: the consumers in C will now lose to the benefit of the producers in B. But the total residual gains will still be measured fully by the change in the area under the demand curve for transport between A and B. With a multiplicity of regions, however, solving for the transport demand function between A and B becomes more complicated.

Variable Transport Costs

In the last two sections, it was assumed that the unit transport costs on the roads connecting the regions are constant over the range of traffic considered. In some cases, especially where rural areas are concerned, this may be a realistic assumption. However, the marginal transport cost per unit of commodity will tend to rise as the volume of traffic increases, at least beyond a certain traffic density.

The most important cause of a rising road transport supply curve is, of course, congestion. In the absence of a proper congestion tax, congestion will create a divergence between social and private marginal cost of transport.[12] Such divergences between social and private costs raise special questions which will be discussed in Chapter IV. The present section assumes, unrealistically, that such divergences are absent and concentrates only on the implications of a rising supply curve of transport services.

We first assume that in our three-region model only the transport supply curve on the road between the two regions A and B is rising. The transport cost on the road between A and the competing region C is assumed constant as in the last section. The improvement of the road between A and B will shift the supply curve downward, i.e. the cost of transporting a given volume of traffic will be lower. We shall assume that the new supply curve will be parallel to the pre-improvement one.

In this case, the total benefits to consumers and producers in the regions A, B, and C will be given by the same expressions as before:

$$\underline{\Delta} P_A(T_{AB}{}^0 + \tfrac{1}{2}\Delta T_{AB}) + \Delta P_B(T_{AB}{}^0 + \tfrac{1}{2}\Delta T_{AB})$$

or

$$\underline{\Delta}\theta_{AB}(T_{AB}{}^0 + \tfrac{1}{2}\Delta T_{AB})$$

[12] See A. A. Walters, *The Economics of Road User Charges*, World Bank Staff Occasional Paper No. 5 (Baltimore: Johns Hopkins Press, 1968), p. 25.

except that the determination of the new equilibrium values of prices, quantities and transport costs must take into account the rising transport cost function. The changes in the transport cost to shippers, in prices, and in the volume of traffic will be smaller in this case than if the transport supply function were horizontal.

This measure is, however, incomplete. It should be modified to take into account the change in the surplus accruing to the producers of the transport services, as illustrated in Figure 3. The horizontally shaded area *FACE* equals the total of benefits to the producers and consumers of the commodity in the three regions, i.e. $\Delta\theta_{AB}(T_{AB}{}^0 + \frac{1}{2}\Delta T_{AB})$. *FE* is the fall in the unit transport cost ($\Delta\theta_{AB}$), and $T' - T^0$ is the increase in traffic (ΔT_{AB}). Both of these are smaller than they would have been if the new supply curve *QS′* had been horizontal, i.e. *KS″*. With curve *KS″*, the reduction in the transport cost for the initial volume of traffic T^0 would have been the same (i.e. the downward shift at T^0 is the same), but the transport cost to consumers would have fallen more, to *K* rather than *E*, and the volume of traffic would have increased more, to *KC′* rather than *EC*. The benefits to the consumers of the road transport services would then have been greater, *FAC′K* rather than *FACE*.

What happens to the producers of these services? They were earning a surplus of *FAP* with the pre-improvement curve PS^0. With the post-improvement supply curve *QS′*, the surplus is *ECQ*, which is greater than

FIGURE 3
VARIABLE TRANSPORT COST ON ROAD FROM A TO B

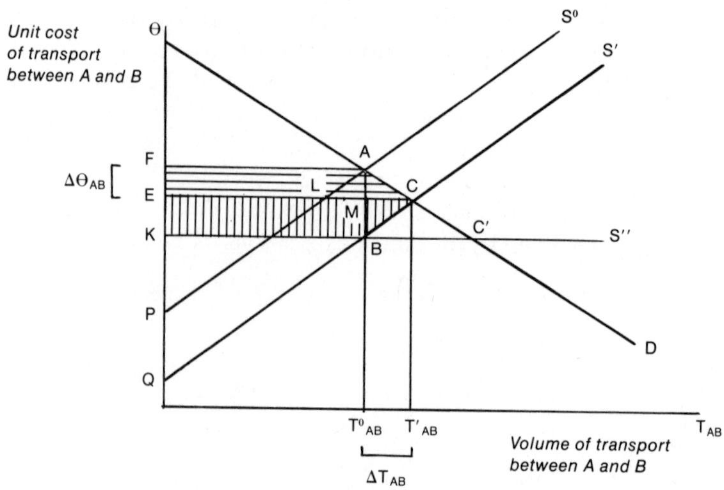

the old surplus by the amount of *ECBK*, the vertically shaded area.[13] Thus the transport industry makes an additional profit of *ECBK* due to the improvement of the road, which should be added to the benefits to the consumers and producers in the three regions, *FACE*, to get the total benefits attributable to the road project, *FACBK*. Even allowing for this increase in profit by the transport industry, total benefits from the road improvement, *FACBK*, are smaller than they would have been with a horizontal transport supply curve, *FAC'K*.

This suggests an extension of the previous benefit measure.[14] If s is the vertical shift in the transport supply curve (*AB* in Figure 3), then the total benefits due to the road project can be measured by $s(T_{AB}^0 + \frac{1}{2}\Delta T_{AB})$. In the case of a rising supply curve, s will be greater than the actual change observed in the transport cost $\Delta\theta_{AB}$. The total benefits will thus equal the increase in the area between the demand and supply curves for transport, and not just the increase in the area below the demand curve as before.

Let us consider next what happens if the transport supply curve between the regions *C* and *A* is also rising. It turns out that the measure given above for the benefits from an improvement of the road between *A* and *B*, i.e. $s(T_{AB}^0 + \frac{1}{2}\Delta T_{AB})$, remains valid. But the increase in traffic between *A* and *B*, i.e. ΔT_{AB}, can no longer be interpreted as a movement along a transport demand curve between *A* and *B*, because as the transport cost on the road between *A* and *C* changes, the demand curve for transport between *A* and *B* will be shifting. Hence, it will not be meaningful to speak of the *ceteris paribus* demand for transport between *A* and *B*. The change in the volume, ΔT_{AB}, can, however, be interpreted as a movement along a locus of price-quantity equilibrium points, defined by the intersections of the

---

[13] To see this, note that:

$$\begin{aligned} ECQ - FAP &= LCQP - FALE \\ &= PABQ - ALM + BMC - FALE \\ &= ABKF - FALE - ALM + BMC \\ &= ALM - ALM + BMC + EMBK \\ &= BMC + EMBK \\ &= ECBK. \end{aligned}$$

[14] The results hold not only for the simple constant-slope cost curve shown, but also for the more usual road transport cost curves with an initial horizontal segment and an upward sloping segment for higher volumes of transport. The only condition is that the cost curve shifts downward by the same absolute amount over the whole relevant range of transport volumes. If the cost curve shifts "to the right"—if the road improvement increases capacity rather than lowering operating cost—the extended measure obviously cannot be used. The benefit to consumers of road services is measured as before by $\underline{\Delta\theta}(T_{AB}^0 + \frac{1}{2}\Delta T_{AB})$ but the reader may easily verify that the producers of these services may then gain or lose, on balance.

shifting demand curve between A and B with the new transport supply curves between A and B.

As the road between A and B is improved and the unit transport cost at any given level of traffic between A and B falls, the demand curve for transport between A and C will shift to the left, transport cost between A and C will fall, and the demand curve for transport between A and B will also shift to the left. In the new equilibrium, we will be likely to have a lower θ_{AB}, a lower θ_{AC}, a lower T_{AC}, and a higher T_{AB}. This is illustrated in Figure 4, where $(\theta_{AB}^0, \theta_{AC}^0, T_{AB}^0, T_{AC}^0)$ is the set of values in the initial equilibrium. After S_{AB}^0, the transport supply curve between A and B, falls to S_{AB}' because of the road improvement, $(\theta_{AB}', \theta_{AC}', T_{AB}', T_{AC}')$ becomes the set of values in the new equilibrium. Both the demand curves have shifted to the left. θ_{AB} and θ_{AC} are lower than before; the traffic between A and C has fallen, and traffic between A and B has increased.

From these two equilibria, we might imagine a series of shifts in S_{AB} giving rise to a series of new equilibrium values (θ_{AB}, T_{AB}), shown by the locus LL. The locus of all the equilibrium values of (θ_{AC}, T_{AC}), S_{AC}^0, does not shift because the road between A and C is not improved.

Turning our attention to the measure of the benefits from the road improvement between A and B, our difficulty in redefining the demand curve for transport between A and B does not affect the basic measure of the gain to the consumers and producers of the commodity in the three regions. As before it is given by:

$$\underline{\Delta}P_A(T_{AB}^0 + \tfrac{1}{2}\Delta T_{AB}) + \underline{\Delta}P_A(T_{AC}^0 - \tfrac{1}{2}\Delta T_{AC}) + \Delta P_B(T_{AB}^0 + \tfrac{1}{2}\Delta T_{AB}) - \underline{\Delta}P_C(T_{AC}^0 - \tfrac{1}{2}\Delta T_{AC})$$

FIGURE 4
VARIABLE TRANSPORT COSTS ON
ROAD FROM A TO B AND FROM A TO C

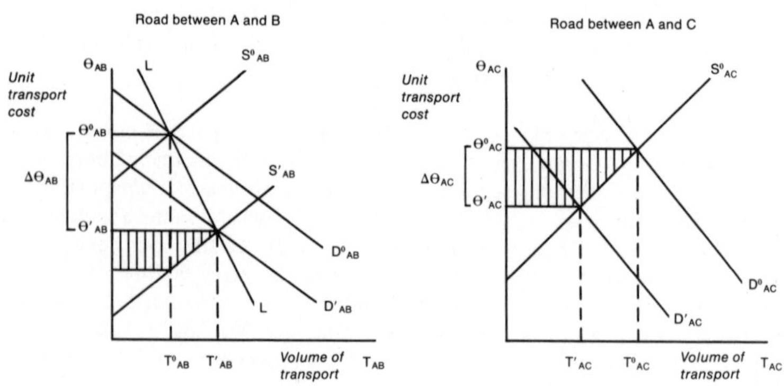

where the first two terms measure the residual gain in A, and the following two terms the residual gain in B and loss in C, respectively. But unlike our earlier cases, the fall in price in A does not equal that in C, because transport costs between A and C decrease instead of remaining constant as before. Hence, the benefits are in this case:

$$(\underline{\Delta}P_A + \Delta P_B)(T_{AB}{}^0 + \tfrac{1}{2}\Delta T_{AB}) + (\underline{\Delta}P_A - \underline{\Delta}P_C)(T_{AC}{}^0 - \tfrac{1}{2}\Delta T_{AC})$$
$$= \underline{\Delta}\theta_{AB}(T_{AB}{}^0 + \tfrac{1}{2}\Delta T_{AB}) + \underline{\Delta}\theta_{AC}(T_{AC}{}^0 - \tfrac{1}{2}\Delta T_{AC})$$

The second term did not previously appear in the expression since, when transport costs between A and C are constant, $\Delta\theta_{AC}$ equals zero and ΔP_A equals ΔP_C.

Even this measure is not yet complete. It does not take into account the changes in producers' surplus of the transport industry between A and C and between A and B. The loss to the producers of the transport services between A and C is $\underline{\Delta}\theta_{AC}(T_{AC}{}^0 - \tfrac{1}{2}\Delta T_{AC})$, i.e. the vertically shaded area (for the road between A and C) in Figure 4. Deducting this loss reduces the total gain to the familiar expression: $\Delta\theta_{AB}(T_{AB}{}^0 + \tfrac{1}{2}\Delta T_{AB})$. And after adding the gain to the producers of the transport service between A and B, the vertically shaded area (for the road between A and B) in Figure 4,[15] the total benefits from the road improvement become once more $s(T_{AB}{}^0 + \tfrac{1}{2}\Delta T_{AB})$, where s, as before, is the downward shift of the transport supply curve between A and B.

To summarize the arguments of this chapter:[16] if T^0 is the normal volume of traffic between A and B, ΔT is the increase in the volume of transport between A and B because of the road improvement between A and B, s is the downward shift in the supply curve of transport between A and B, and $\Delta\theta$ is the change in transport cost to the shippers between A and B, then the total residual gain, taking account of gains and losses not only to consumers and producers of the commodity in the three regions but also to producers of the transport services between A and B and between A and C, is given by $s(T^0 + \tfrac{1}{2}\Delta T)$. In other words, the total residual gain is measured by the increase in the area between the demand curve (or locus) and the supply curve for transport between A and B. The reader will recall that this measure of the benefits of a road improvement was derived on the assumption that income distribution and external effects may be ignored, that pricing is based on marginal social cost, and that the road improve-

[15] This corresponds to the vertically shaded area $ECBK$ in Figure 3. The gain to the producers of the transport service between A and B is measured by

$$(s - \underline{\Delta}\theta_{AB})(T_{AB}{}^0 + \tfrac{1}{2}\Delta T_{AB})$$

[16] For simplicity subscripts have been dropped.

ment results in a uniform downward shift in the transport supply curve over the whole relevant range.

If the transport costs on both roads remain constant whatever the volume of traffic, then s equals $\underline{\Delta}\theta$ and ΔT is a movement along the demand curve for transport between A and B. If transport costs are increasing with traffic on road AC but are constant on road AB, then s still equals $\underline{\Delta}\theta$, but ΔT is now a movement along the price-quantity locus of demand for transport between A and B. If transport costs are increasing with traffic on road AB, but constant on road AC, then s is greater than $\underline{\Delta}\theta$ and ΔT is a movement along the demand curve for transport between A and B. Finally, if transport costs are rising with traffic on both roads, then s is greater than $\underline{\Delta}\theta$ and ΔT is a movement along the price-quantity locus. In other words, a positively sloping supply curve for transport between A and B results in the shift in the supply curve exceeding the drop in unit transport cost; a positively sloped supply curve for transport between A and C requires that the demand curve for transport between A and B be replaced by a different price-quantity locus.

These results indicate that as long as the new equilibrium volume of transport between A and B can be estimated, and as long as the vertical shift in the transport supply curve between A and B can be estimated, the benefits can be measured quite easily. It does not matter whether there are competing regions or whether the transport supply curve between A and B and the transport supply curves between A and other regions are positively sloped. These modifications affect only the estimation problem. With a horizontal transport supply curve, we need to know only the shift in the supply curve and the elasticity of the demand curve for transport between A and B. With a rising transport supply curve between A and B, but a horizontal one between A and C, we need to know also the supply elasticity for transport between A and B. With a positively sloped supply curve between A and C as well, the interdependence effect between the transport demand curves must also be known.

III

MEASURING BENEFITS IN THE PRESENCE OF COMPETING TRANSPORT

When we introduce competing routes into the model of Chapter II, we open the possibility that traffic from these other routes may also benefit from the improvement of the road by switching to the improved road. We call such traffic *diverted traffic*. We did not encounter this kind of traffic in Chapter II because we assumed that the road under improvement was the only available route from *A* to *B*.

It is true that in Chapter II the improvement of the road between *A* and *B* in the three-region model did improve the trading position of *B* relative to *C* and did create benefits for *B* at the expense of *C*, but in the absence of competing modes or routes this kind of traffic diversion worked completely through the market for the commodity transported in trade. We showed that, in the absence of market imperfections and ignoring changes in income distribution, the gains and losses of such trade diversion are fully accounted for by the sum of road user savings and induced benefits, measured by the increase in the area between the supply and demand curves for transport on the improved road. This in turn depends on supply and demand elasticities for the commodities transported, on the reduction in transport costs and on transport and supply elasticities. Losses due to the possible decline of traffic on other sections of the network are already implicit in the demand curve for transport on the improved road and need not be separately accounted for.

We will see in this chapter to what extent the measures used in Chapter II need to be modified in order to arrive at total benefits when traffic is diverted from competing routes. Let us suppose that apart from the road to be improved, there is another way of traveling from *A* to *B*. We want to know in what way the measurement of benefit already developed needs to be modified to take into account the fact that some traffic on the competing route will switch to the improved road. Before taking up the more general case, we first consider the problem under the assumption that the two competing transport facilities provide the same service.

When Competing Services Are Perfect Substitutes

As far as traffic in commodities is concerned, the analysis of traffic diversion is almost always based on the assumption that the competing transport facilities provide the same service, e.g. "transportation" between *A* and *B*.[1] This assumption, which is fundamental to transport network models in particular, has a very simple consequence. If, for example, there are two competing roads between two regions, *A* and *B*, then either the marginal transport cost will be the same on both roads or only one of them will be used.

In other words, the assumption means that a shipper sending a ton of wheat, say, from *A* to *B*, will be indifferent between the two roads as long as the cost of shipping (including the costs of vehicle operation, safety factors, transit time, etc.) is the same on both roads. This assumption, of course, may not be realistic in all cases (particularly for passenger traffic) and the implications of suspending it will be discussed in the next section of this chapter.

Under this assumption, the problem becomes the simple one of determining the allocation of total supply between two competing suppliers of the same service. The relevant supply curve will be the sum of the separate supply curves for the services of the two roads. With upward sloping supply curves, an improvement of the road will mean a downward shift of the total supply curve. The area under the transport demand curve will increase as will the producers' surplus; this will be partially offset by the loss of producers' surplus on the unimproved road. Total gains will be measured by the change in the area between the total demand curve and the total supply curve of transport between regions *A* and *B*, as is shown in Figure 5.

In the figure, S_1^0 and S_2^0 are the respective supply curves of the two roads' services between regions *A* and *B* without the improvement. S_1' is

[1] Actually, all that is required for this case is that the two services be perfect substitutes over the range of price changes under consideration.

FIGURE 5
DEMAND AND SUPPLY OF TRANSPORT
COMPETING MODES PERFECT SUBSTITUTES

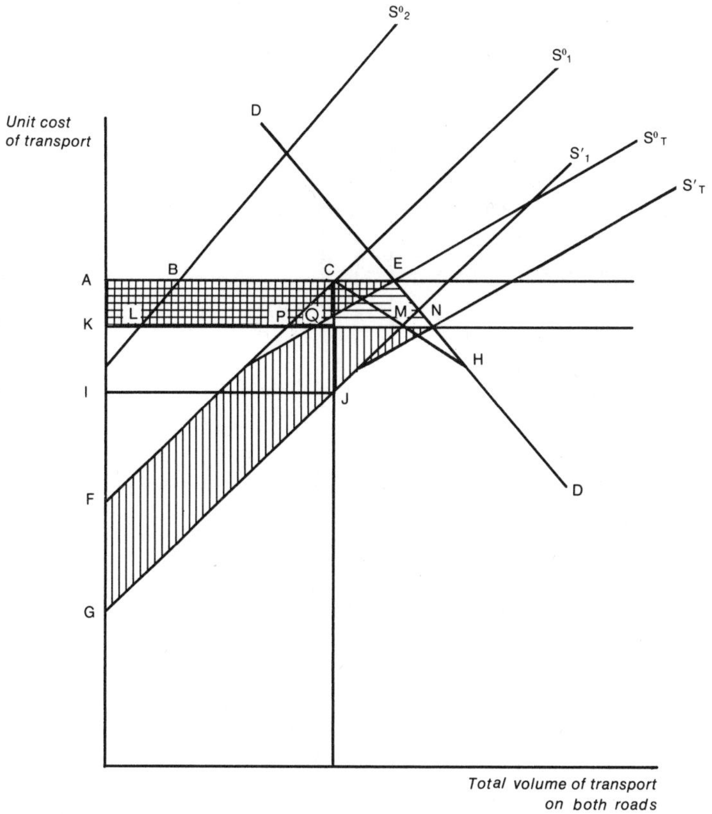

the supply curve of the project road's services after its improvement. $S_T{}^0$ and $S_T{}'$ are the total supply curves of transport between regions *A* and *B* without and with the improvement. *DD* is the demand curve for transportation services between the two regions. Initially, total supply and demand equal *AE*, of which *AB* is supplied by the second road and *BE* = *AC* by the project road to be improved.

With the improvement on the first road, traffic on the other road declines to *KL* and traffic on the improved road rises to *LN* = *KM*. The gain to road users is measured by the horizontally shaded area *AENK*, and the producers of transportation services between *A* and *B* gain the vertically shaded area but lose the cross-hatched area. Total gains, therefore, amount to the areas shaded horizontally and vertically, *FENG*, i.e. the band below

FIGURE 6
MEASURING BENEFITS
IN THE PRESENCE OF TRAFFIC DIVERSION

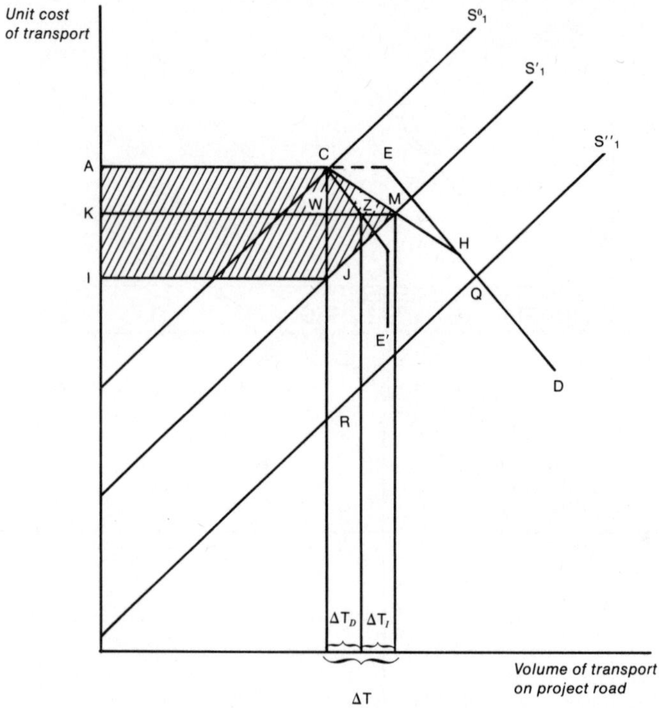

the demand curve and between the total supply curves without and with the improvement.

By construction $KL = MN = PQ$. As a result the area of the triangle *QEN* is equal to the area of the triangle *PCM*. The producers' surplus on the improved road, *FQNG*, is, of course, the same as *FPMG*. Consequently the area *FENG* equals the area *FCMG*. This equals the area $ACJI + CJM$, or $s(T^0 + \frac{1}{2}\Delta T)$, where s is the uniform downward shift in the supply curve of the first road's services.[2] All this is more clearly seen in terms of Figure 6 above, derived from Figure 5.

In Figure 6 the demand curve for the improved road's services *CMHD* is an excess demand curve—the excess of total demand over the amount supplied by the other road. As a consequence, up to the point *H* it is more elastic than the *DD* curve in Figure 5, after which it becomes identical

[2] For the same reasons as in Figure 3. T^0 and ΔT here refer to traffic levels on the project road, i.e. $T^0 = AC$ and $\Delta T = LN - AC$.

with DD. The point H represents the unit cost at which all traffic from the other road is diverted to the improved road. The two supply curves, S_1^0 and S_1', together with this excess demand curve determine the initial and final equilibrium traffic levels and costs. Initially AC was the amount of traffic on the project road (as in Figure 5) and after its improvement it gains an amount of traffic WZ (equal to AB-KL in Figure 5) at the expense of the unimproved road and an amount ZM of newly generated traffic. The benefit measure is now $ACMJI$, which is identical to $ACMJI$ of Figure 5. This equals $s(T^0 + \frac{1}{2}\Delta T)$, the traffic levels being measured on the project road.

With reference to Figure 6, we now look at the curve CE', which would be the excess demand curve for the improved road's services if we assumed that the total demand for transport were price inelastic (i.e. that curve DD in Figure 5 were vertical). This curve CZE', shows the increases in traffic on the improved road due to traffic diversion only. That is, in the pure diversion case CZE' would have been the relevant excess demand curve for the improved road's services. Note that ΔT equals $\Delta T_D + \Delta T_I$, so that we can write the total benefits as $sT^0 + \frac{1}{2}s\Delta T_D + \frac{1}{2}s\Delta T_I$. With pure diversion ΔT_D would be larger and ΔT_I would be zero. Similarly, in the case when there is no traffic diversion (no roads which are perfect substitutes for the road being improved), ΔT_D equals zero.

Thus, in this example, the measure $sT^0 + \frac{1}{2}s\Delta T$, where the traffic levels refer to the project road, is applicable to the pure diversion case ($\Delta T_I = 0$), the pure induced traffic case ($\Delta T_D = O$) as well as to the mixed case. However, suppose that the post-improvement supply curve was S_1'' instead of S_1'. In this case all traffic on the other road would have completely switched to the improved road as S_1'' cuts the demand curve below point H. The excess demand curve for transport on the project road is then kinked in the relevant range. Instead of a triangle like CMJ we get the area $CHQR$ as part of the benefit measure; this will be larger than the triangle formed by C, Q and R, i.e. larger than $\frac{1}{2}s\Delta T$. The measure will in such cases of complete diversion underestimate the actual benefits.

Whereas the principle of benefit measurement, viz. gain to road users *plus* producers' surplus gain on the project road *minus* loss of producers' surplus on the other road remains the same, the form of the benefit measure will be quite different from $s(T_0 + \frac{1}{2}\Delta T)$ in the presence of diversion, even assuming that the assumption of a uniform downward shift in the project road's supply curve is valid. Complete diversion from alternative routes may be fairly common. This suggests that, where such diversion is a major issue, it is preferable to analyze the changes due to a road project in terms of the transport market as a whole, as in Figure 5.

When Competing Services Are Imperfect Substitutes

The analysis in the previous section was based on the assumption that the competing roads (or transport modes) provide the same service. If this assumption is not justified then we no longer have the traffic diversion case, strictly speaking. The problem is no longer that of determining the allocation between competing suppliers of the same service, but of determining the impact of a change in the market for one product on the markets for other related products. In other words, the services of the project road and those of another road or transport mode will have mutually dependent but separate demand curves. The analysis of benefits in this case may be quite complex, as indicated in the Annex. However, under some assumptions the benefit analysis becomes very simple and the benefit measure takes the familiar form. We shall only discuss the simple case here, leaving for the interested reader the more general formulations in the Annex.

Suppose that the supply curve for the services of the other road (or mode) is positively sloped.[3] This implies that the cost of the other service will fall when its demand curve shifts to the left due to the cost reduction on the project road. We will have the normal interaction effects: this fall in cost will shift the project road's demand curve to the left. If the new supply curve on the project road is also positively sloped, there will be a further fall in costs on the project road which will, in turn, affect the demand curve for all the other services, and so on, until the final equilibrium is reached. In Figure 7, (θ_R^0, T_R^0) and (θ_R', T_R') are the price and quantity of road transport without and with the road improvement, and (θ_L^0, T_L^0) and (θ_L', T_L') are the corresponding prices and volumes for the other service. The gain in terms of consumers' surplus for road users and for the users of the other service is then measured by the sum of the horizontally shaded areas.[4] Consumers of both the road service and the other service gain as prices for both services decline. Producers of the other service suffer a loss of surplus which exactly offsets the gain in consumers' surplus of the users of that service. In addition there is a net gain in producers' surplus on the project road, measured by the vertically shaded area. The residual benefits

[3] We assume, as we do throughout this chapter, that the supply curve represents the marginal social transport cost curve, there being no difference between marginal social and private costs.

[4] For this to be strictly correct the demand functions must be linear, income effects must be nil, and the cross-partial derivatives must be constant. If the cross-partial derivatives are not constant (e.g. if the effect on demand for road services of a unit change in the price of other services depends on the initial price of road services), then a more generalized measure of benefits has to be used. This case is discussed in the Annex.

FIGURE 7
MEASURING BENEFITS
MARKET INTERACTIONS

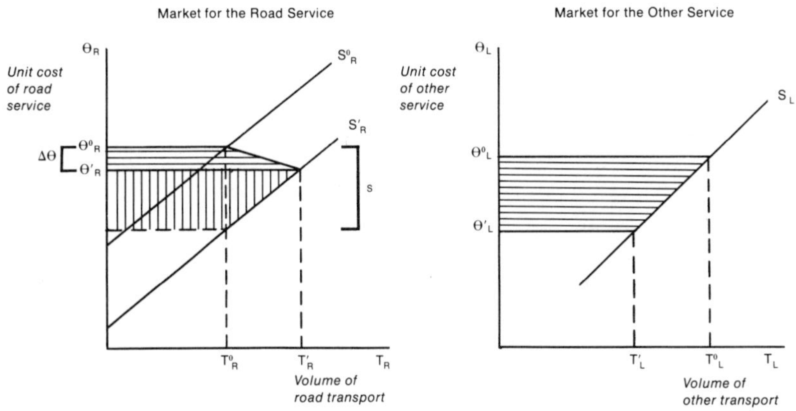

of the road improvement are therefore measured by the sum of the horizontally and vertically shaded areas for road alone. If s is again the uniform downward shift in the supply curve for road services, the benefits are measured by the familiar expression $s(T^0 + \frac{1}{2}\Delta T)$, where T^0 is the normal volume of road transport ($T_R{}^0$ in Figure 7) and ΔT the increase in road transport ($T_R{}' - T_R{}^0$).

If the price of the other service remains constant when the road is improved, there will be no gain of consumers' surplus nor offsetting loss of producers' surplus on that service. The measurement of benefits remains the same, $s(T^0 + \frac{1}{2}\Delta T)$, but the set (θ_R, T_R) in that case traces out a demand curve, with other prices held constant, and not a locus of transport allocation equilibrium points as in Figure 7. Finally, if the road transport supply curve is horizontal, then s equals $\underline{\Delta\theta}$ which in turn equals $\theta_R{}^0 - \theta_R{}'$, i.e. the benefits are fully measured by the increase in consumers' surplus on road services.

These results are analogous to those arrived at in the three-region case of the previous chapter: the benefits are measured by the change in area between the demand and supply curves for the improved road, and the decline in traffic on competing roads does not have to be separately taken into account. In fact Figures 4 and 7 are similar. In the three-region case the demand for transport and the benefit measure were derived from the demand and supply conditions for the commodity in the various regions. In the present case of diversion between competing services on the same route, the joint demand functions for these services are specified independently. The benefit measure is derived directly from a utility function for these services, not indirectly from changes in consumers' and producers' surpluses on the transported commodity.

The results in this and the foregoing sections have assumed, rather unrealistically, that external effects are absent. Important qualifications are necessary if these assumptions are dropped; this issue will be discussed in the next chapter.

IV

THE EFFECTS OF MARKET IMPERFECTIONS

The discussion thus far has assumed that transport costs are the only source of price differentials between regions, and that the transport rates are based on marginal costs to the economy and reflect the marginal value of transport services to the economy. In reality, transport rates may be higher than marginal cost because of monopolistic rate setting, cross-subsidization between different routes and products, government interventions, road user charges, etc. Such market imperfections are common. In this section we consider various forms of market imperfections and show that they may call for substantial changes in our previous measures of the benefits from a road improvement.

Inequality of Price and Marginal Cost

Beginning with the assumption that marginal private and social costs are equal, we consider the implications of a pricing policy that keeps the price level above the marginal cost, as illustrated in Figure 8. θ^0 and θ' are the initial and new prices (transport costs to users) respectively. M^0 is the marginal cost[1] of supplying T^0, and M' is the marginal cost of supplying T'.

[1] In this chapter, we distinguish between θ, the actual price charged, and M, the marginal transport cost. In previous chapters, we have assumed θ equal to M and have called θ the unit transport cost.

FIGURE 8
MEASURING BENEFITS
PRICE GREATER THAN MARGINAL COST

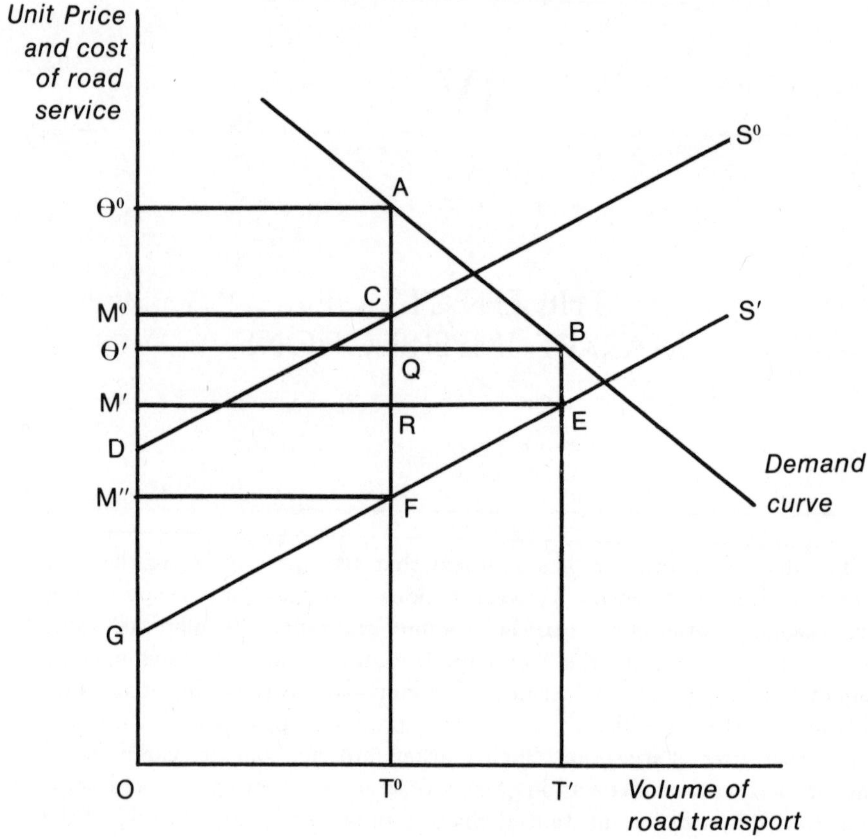

The road improvement is represented by the downward shift of the supply curve, S^0 to S'. At the initial level of transport, for example, the marginal cost falls to M'' from M^0.

The benefit to road users in this case is measured as before: $\Delta\theta(T^0 + \frac{1}{2}\Delta T)$, represented in Figure 8 by the area $\theta^0 AB\theta'$. The change in producers' surplus, however, is now different. On the initial traffic, T^0, producers' surplus changes by the area $\theta^0 ACD - \theta' QFG$, i.e. by $T^0(s - \Delta\theta)$—where s is, as before, the shift (assumed uniform) in the supply curve. On the induced traffic there is a further change, measured by the area $QBEF$, or $m'\Delta T + \frac{1}{2}d\Delta T$—where m' denotes the difference between price and marginal cost in the new situation, and d denotes the difference between the marginal costs of producing T' and T^0 in the new situation. The total increase in benefits (shown by the area $M^0CFM'' + ABEF$) is equal to the sum of all the above elements, i.e. to $\underline{\Delta}\theta(T^0 + \frac{1}{2}\Delta T) + T^0(s - \underline{\Delta}\theta) + m'\Delta T + \frac{1}{2}d\Delta T = sT^0 + \frac{1}{2}\underline{\Delta}\theta\Delta T + \frac{1}{2}d\Delta T + m'\Delta T$.

Notice that this discussion did not take into account how the price marginal cost differences are actually determined. For example, if the distortions are due to an ad valorem tax on road services, m^0 and m' will equal τM^0 and $\tau M'$, τ being the tax rate. In this case, a part of the change in producers' surplus, as measured in the previous paragraph, will represent a change in the tax revenue, which may of course be positive or negative. One can similarly analyze monopolistic rate setting, although the diagrammatic representation would be different.

Thus, this discussion of the effects of market imperfections applies to road user taxes, which often raise transport rates above the social marginal cost. The implications of such tax distortions can be analyzed along the above lines. When the distortions are due to a monopolist, however, the changes in his monopoly profits are taken into account, while in the case of a tax, it is the losses (gains) of tax revenue which must be allowed for.

The same considerations apply to cases where the price differential between two regions is kept above the marginal *social* cost of transport (including distribution). For example, the trucking industry may be perfectly competitive and pass cost reductions on to shippers, but if this results in higher monopoly profits for middlemen, the analysis given above still applies. The only difference is that it is the middlemen's rather than the truckers' gains and losses of monopoly profits which we consider.

Instead of a comprehensive taxonomy of the many possible cases that can arise due to road user taxes, toll rates, monopolistic markets in road services or in distribution, it should suffice to state the basic modifications that will be necessary to the "competitive" measure in all these cases. Changes in monopoly profits or in tax revenue should be explicitly taken into account, having regard for the fact that changes in tax revenue may need to be evaluated differently from monopoly profits. Moreover, in all these cases differences between marginal private and social costs would present a further complication. In the following section we shall consider one common reason for such differences between private and social costs, assuming away all other complications.

Congestion

When there is congestion, the marginal social cost will exceed the marginal private cost, contrary to our assumption thus far. A road user will take account of the effect of congestion on his own private cost, but not of the increase in congestion cost he causes to other users. Since offsetting congestion charges are seldom used, marginal social cost will then exceed marginal private cost, i.e. average social cost. Assuming demand to be based on marginal private cost, i.e. no additional market imperfections of the kind discussed above, the level of traffic both with and without the road improvement will exceed the socially optimal level.

Figure 9 shows how the measure of benefits needs to be corrected to allow for this case. In this figure, SS^0 and SS' are the social marginal cost curves without and with the road improvement. T_s^0, T_s', ΔT_s and s are the corresponding volumes, change in volume and the downward shift in the social supply curve. The subscript s refers to the fact that these values correspond to the social marginal cost curves, SS^0 and SS', and the subscript p refers to the private marginal cost curves, PS^0 and PS'.

FIGURE 9
MEASURING BENEFITS
CONGESTION

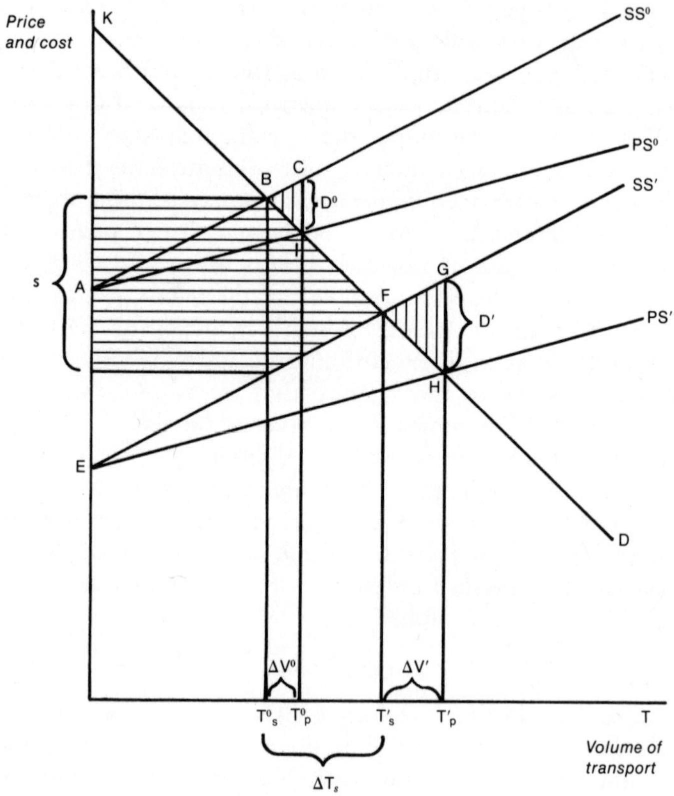

The benefits to the consumers and producers of transport services are then not measured by the horizontally shaded area $s(T_s^0 + \frac{1}{2}\Delta T_s)$, as this measure is not corrected for the overexpansion of traffic volume both before and after the road improvement. The cost to society of the excess traffic without the improvement is measured by the vertically shaded triangle $\frac{1}{2}D^0\ \Delta V^0$ where D^0 is the excess of the marginal social cost over the marginal private cost of producing T_p^0, and ΔV^0 is the excess traffic $T_p^0 - T_s^0$. Similarly, the cost to society of the excess traffic with the improvement is measured by the vertically shaded triangle $\frac{1}{2}D'\ \Delta V'$. The road improvement eliminates the loss on the initial excess traffic but entails a new loss on the excess traffic after the improvement.

Total benefits are therefore: $s(T_s^0 + \frac{1}{2}\Delta T_s) + \frac{1}{2}D^0\ \Delta V^0 - \frac{1}{2}D'\ \Delta V'$.[2] If one may assume parallel shifts in linear supply curves, then $\frac{1}{2}D'\ V'$ is greater than $\frac{1}{2}D^0\ \Delta V^0$, and therefore the divergence between marginal social and private costs which results from congestion will reduce the benefits of a road improvement.

There are many variants of this case, determined by the shape and shift of the cost curve. Consider, for example, Figure 10, in which the existing transport cost curve is drawn more realistically, with a horizontal segment and an upward curving segment. In this figure the road improvement eliminates all congestion, so that the lower cost curve is horizontal over the relevant range. In the absence of a congestion tax, traffic before the improvement will be T_p^0. The excess traffic creates a social cost, measured as before by the vertically shaded area. The road improvement not only eliminates the social cost of this excess traffic, but also provides net benefits, measured by the horizontally shaded area. Total benefits are therefore *ABCDEF* in this case.[3]

This measure assumes, of course, that it is impossible for administrative or political reasons to eliminate the existing excess traffic directly by im-

[2] Perhaps the easiest way of showing this is as follows: the difference between the willingness to pay for the initial volume of traffic, T_p^0, and its total social cost is the difference between the areas under *KI* and *AC*, which equals *KBA* minus *BCI*. Similarly, the willingness to pay for traffic T_p' minus the total social cost of T_p' is *KFE* minus *FGH*. The increase in willingness to pay minus the increase in total social cost is then *KFE* – *KBA* – (*FGH* – *BCI*), which equals area *ABFE* + *BCI* – *FGH*. By construction, area *ABFE* equals $sT_s^0 + \frac{1}{2}s\Delta T_s$. *BCI* and *FGH* are of course, $\frac{1}{2}D^0\ \Delta V^0$ and $\frac{1}{2}D'\ \Delta V'$ respectively. Hence, the total benefit from the change is measured by the equation in the text.

[3] Because of the marginal/average relationship of SS^0 and PS^0, the area *KDEF* is an equivalent measure of total benefits. This is contrary to M. E. Beesley and A. A. Walters, "Some Problems in the Evaluation of Urban Road Investments," *Applied Economics*, Vol. 1, No. 4 (1970), p. 248.

FIGURE 10
MEASURING BENEFITS
INCREASED TRANSPORT CAPACITY

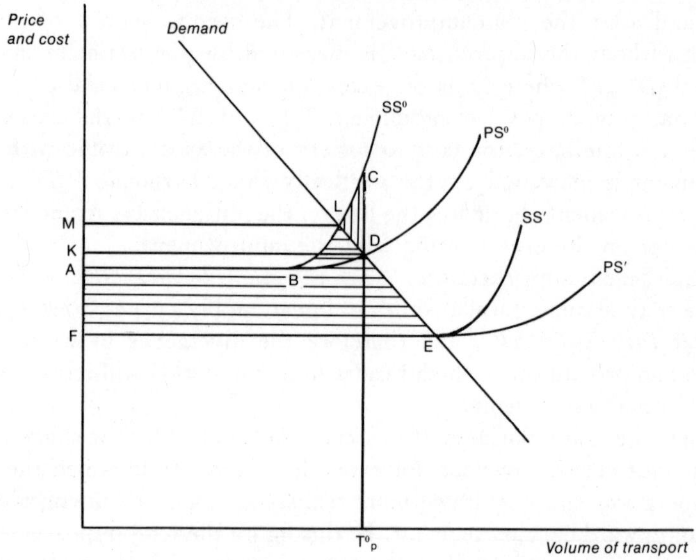

posing a congestion tax. In other words, the measure of the benefits takes the existing non-optimum pricing policies as given. If a policy of congestion pricing were feasible, the benefits from the road improvement would, of course, be only *ABLEF*, as discussed in earlier sections of this paper.[4]

Competing Transport and Market Imperfections

Divergences between prices and social costs of transport also often exist on competing roads or railways. Diversion of traffic when the road is improved will then give rise to other gains or losses not included in the measures of benefits we have presented so far. For example, if rail rates exceed marginal social cost (assumed equal to marginal private cost) because of monopolistic pricing or taxation, a contraction in rail traffic will entail an additional loss.

This is illustrated in Figure 11. After the improvement of the road, an amount of traffic, L^0L', shifts from rail to the road ($L^0L' < R^0R'$). The incremental social surplus on road is measured, as before, by the shaded area on the left-hand side. A further question now is how the railway

[4] The standard formula $s(T^0 + \frac{1}{2}\Delta T)$ does not apply of course in this case. Here the supply curve was not shifted uniformly downward, but "to the right."

FIGURE 11
MEASURING BENEFITS
PRICE DISTORTION ON RAILWAYS

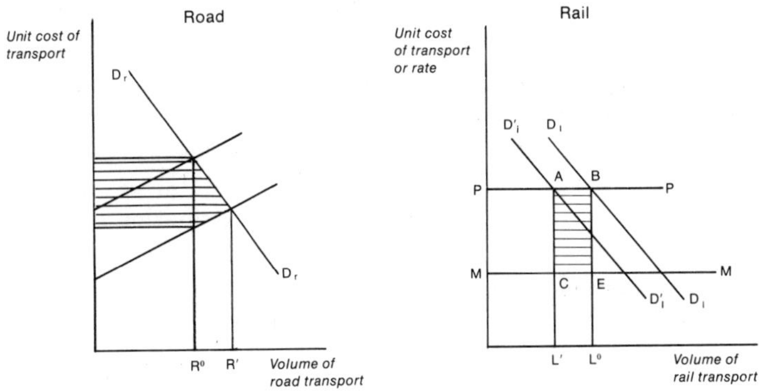

rate is related to the marginal cost on railways. The definition of the latter depends on the kind of effect the transfer of traffic has on the operation of the railways. If only an intermediate link of the railway is losing traffic, the size and schedule of trains may be unaffected by the road improvement. In that case the relevant marginal social cost (in the absence of congestion on railways) will be constant and very low—near zero. This is indicated by the *MM* curve on the right-hand side. The rate, however, will almost inevitably be higher than this, given the conventional rate setting policies. As a result the rectangle *ABCE* will represent the loss of surplus on the railways, and this is to be subtracted from the total social surplus gain on the project road. Needless to say, there would have been an additional gain on railways if the marginal cost exceeded the rates. The argument can easily be adapted to take into account different types of rate setting policies—rates declining or increasing with losses of railway traffic. Note also that the equality of marginal social and private costs is not essential to the argument. It is the marginal social cost curve *MM* that is relevant for measuring the offsetting loss on railways (i.e. area *ABCE*) and not the marginal private cost curve—whether or not it lay above or below the curve *MM*. In other words, it is the loss of social as distinct from private surplus on the railways that is relevant.

This simple road-rail competition case illustrates the type of corrections necessary due to market imperfections, not only in other modes but also in other non-transport markets affected by the project. Decreasing congestion on competitive roads and transport modes, and increasing congestion on complementary roads and transport modes are probably the most obvious external effects to be taken into account. The benefits from a road improvement will not, in such cases, be measured correctly if the social surplus measure is applied to the project road only. Additional gains and losses will have to be accounted for.

The foregoing discussion has amply demonstrated that market imperfections give rise to substantial qualifications of our earlier conclusions concerning the measurement of the benefits from a road improvement. It also illustrates that generalizations are more difficult if allowance is made for these complications. Too much depends on the conditions governing any particular case. One clearly should be wary of any mechanical application of the simple formulas given in Chapters II and III of this paper. Even if we do not concern ourselves with income distribution effects and are willing to accept a uniform downward shift in the cost function as a reasonable approximation of the effect of a road improvement on costs, market imperfections tend to be so pervasive that their effects on the measurement of the benefits from a road improvement cannot safely be ignored.

V

CONCLUSIONS

This paper has explored at length the circumstances under which the change in the area between the demand curve for a road's service and its supply curve will fully measure the total benefits realized by improving the road. If the road improvement results in the same absolute unit cost reduction for all levels of traffic within a certain range, and if the demand and supply curves are linear within that range, the total benefits may then be measured by the formula:[1]

$$s(T^0 + \tfrac{1}{2}\Delta T),$$

where

s = the uniform downward shift in the supply curve;
T^0 = volume of traffic on the road without investment; and
ΔT = increase in traffic on the road due to the investment.

A distinction needs to be made, however, between the basic content of the measure and the particular form stated above, which was adopted mainly for expositional ease. This particular form depends on the linearity as-

[1] The traffic volumes must correspond to the same period. For example, in the nth year of the project, T_n^0 will be the normal traffic for that year, i.e. the traffic level expected under the no-investment alternative. ΔT_n will be the increase in traffic over the base T_n^0, during the nth year.

sumptions and the assumption that the supply curve for the project road's services shifts down by a constant amount after the road improvement. The latter assumption may be fairly realistic in cases where the main impact of the road improvement is on operating costs. However, in many cases such an assumption will not be valid—for example, when extra lanes are added to relieve congestion, the quality of the road remaining unaffected. In general, instead of using this measure in the particular form given above, the nature of the supply curve and the way it may be expected to shift should be closely studied before attempting to measure the increase in the area between the demand and supply curves.

Our discussions in Chapters II and III were designed to clarify the content of the measure. We have seen that this measure fully accounts for the gains and losses to the consumers and producers of the transported commodities in the regions affected by the road improvement, as well as for the gains and losses of the suppliers of the road service and of competing transport services. This is a very important property of the social surplus measure as applied to transportation projects and should be fully understood; careful note should be taken of the type of gains and losses under consideration and the assumptions underlying their quantification.

For example, perhaps the most important qualifications to the content of this measure arise from the presence of market imperfections and externalities in the transport industry itself and in related industries. In such cases not all gains and losses will be accounted for by the social surplus measure applied to the project road only. The social surplus measure will have to be extended or modified in ways which cannot be easily summed up in terms of a simple formula. Some illustrations were provided in Chapter IV.

Quite apart from these complications, the social surplus measure, as we have presented it, assumes that the gains and losses of producers and consumers in various regions can be simply added up and subtracted from each other to arrive at the total benefits due to the road project. If such an approach is not acceptable, gains and losses will have to be weighted according to those who benefit and those who lose. Such weighting will also be appropriate to allow for regional preferences, if any.

From a practical point of view the simplicity of the measure, even where legitimate, is deceptive. This measure is to apply only when the level of normal traffic, the extent of cost reductions, and the increase in road traffic due to the cost reduction can be properly estimated.[2] Such an estimation,

[2] Diverted traffic may present additional problems, as discussed on pp. 19–23 in Chapter III.

however, requires knowledge of the various factors that may affect demand and supply of the road's service and of competing services. For this reason, study of the project road alone, without consideration of its place in the wider transport and economic system is unlikely to be satisfactory. Cross-checking the results with those given by other methods of benefit valuation, such as the national income method, may be a useful way to avoid gross error, though care should be taken to avoid mixing methods.[3]

Our remarks in this chapter have indicated that the practical difficulties of estimating benefits to normal and induced traffic arising from a road project are often great. However, our remarks should be understood as qualifications; they do not, we feel, invalidate the undertaking of benefit measurement nor the use of the social surplus method. And though conceptual clarity may not be a sufficient condition for plausible estimation of the benefits of a road improvement, it is still necessary. We hope this paper has increased the reader's understanding of the role in benefit estimation of the factors affecting the demand and supply of road and competing transport services, and has succeeded in clarifying the application of the social surplus concept to the measurement of road benefits.

[3] For instance, once the benefits are measured by the social surplus method, the incremental profits in agriculture or industry should not be regarded as additional benefits to be accounted for. For a useful discussion of the relation between different measures of transport benefits, see Clell G. Harral, *Preparation and Appraisal of Transport Projects*, Washington, D.C.: U.S. Department of Transportation, 1968.

ANNEX

THE MEASUREMENT OF CONSUMERS' SURPLUS

The consumers' surplus measure, as conventionally used, is defined as follows. Suppose the price (x) of a good (1) falls from x^0 to x', the prices ($z_1, z_2, z_3, \ldots, z_n = \bar{z}$) of all other goods remaining constant. Then the incremental consumers' surplus (W) can be written as

$$W = -\int_{x^0}^{x'} X\,dx$$

where $X = \psi(x, \bar{z})$ is the demand for commodity (1) as a function of its own price (x) and of the prices ($\bar{z}$) of other goods, $\bar{z}$ being held constant.[1] This integral represents the area under the demand curve for good (1) between the old and new prices.

If, within the range of the demand function relating to the price change, the function is linear, then the benefit expression becomes:

$$W = |x' - x^0|X^0 + \tfrac{1}{2}(|x' - x^0|)(X' - X^0)$$

[1] This measure depends on the assumption that the marginal utility of income does not change with the change in price. The assumption can also be stated in terms of parallel indifference curves. See P. Samuelson, "Constancy of the Marginal Utility of Income," *Studies in Mathematical Economics and Econometrics*, O. Lange, F. McIntyre and T. O. Yntema (eds.), Chicago: Chicago University Press, 1942; and J. R. Hicks, *A Revision of Demand Theory* (Oxford: Clarendon Press, 1956), Chapter X.

or, with a change in notation ($\underline{\Delta}$ or delta bar is defined as in footnote 4 to Chapter II)

$$W = \underline{\Delta}x\, X^0 + \tfrac{1}{2}\underline{\Delta}x\, \Delta X$$

an expression which the reader has frequently encountered in the text of this paper.

Suppose, however, that when the price of good (1) falls (if the good is road service, this fall may be due to a road investment, for example), the price ($z_1 = y$) of another substitute good (2) falls as well. This is the case discussed in Chapter III of this paper, (pages 24–26). The figure below shows the changes in the market for good (1), and in the market for the competing good (2).

Demand for good (2) is defined as $Y = \phi(x, y, \bar{z})$ and demand for good (1) as $X = \psi(x, y, \bar{z})$, where $\bar{z}$ now $= (z_2, z_3, \ldots, z_n)$ because we are calling $z_1 = y$. To illustrate the shifts, we call ψ^0 the demand for good (1) in the initial equilibrium situation at prices x^0 and y^0, and ϕ^0 the initial equilibrium demand curve for good (2) at prices x^0 and y^0. Likewise, ψ' and ϕ' are the demand functions in the new equilibrium, in which prices are x' and y'. The other prices, $z_2, \ldots, z_n$, are assumed to remain unchanged.

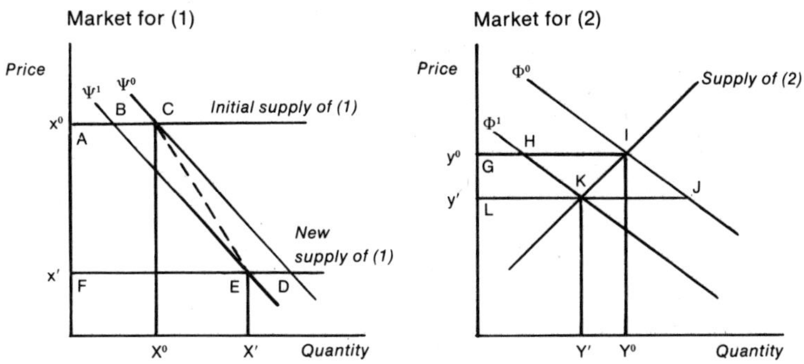

The gain to consumers from the downward shift of the supply curve of (1) can be expressed in either of two equivalent consumers' surplus measures (refer to figure):[2]

Increase in benefit $= W$

(i) $$W = ACDF + GHKL$$

(ii) $$W = ABEF + GIJL$$

[2] See Hicks, *A Revision of Demand Theory* (Oxford: Clarendon Press, 1956), Chapter XVIII, esp. pp. 178–179. This is only true if income effects are zero or negligible.

Since

$$ACDF + GHKL = ABEF + GIJL$$

we have

$$ACDF - ABEF = GIJL - GHKL$$

that is,

$$BCDE = HIJK$$

If the demand curves are linear in the relevant range and if ψ^0 is parallel to ψ' and ϕ^0 is parallel to ϕ', then

$$\tfrac{1}{2}BCDE = \tfrac{1}{2}HIJK$$

implies

$$BCE = IJK$$

We can now simplify the expression (ii) for W.

$$\begin{aligned} ABEF + GIJL &= ABEF + GIJL + BCE - BCE \\ &= ABEF + GIJL + BCE - IJK \\ &= ABEF + BCE + GIJL - IJK \\ &= ACEF + GIKL \end{aligned}$$

This is the measure used in Chapter III of this paper.

In algebraic terms, the measure is simply:

$$\text{(iii)} \qquad \underline{\Delta}x\, X^0 + \tfrac{1}{2}\underline{\Delta}x\, \Delta X + \underline{\Delta}y\, Y^0 - \tfrac{1}{2}\underline{\Delta}y\, \Delta Y$$

where

$$\begin{aligned} \Delta x &= x' - x^0 \text{ (new minus old equilibrium price)} \\ \Delta y &= y' - y^0 \text{ (new minus old equilibrium price)} \\ \Delta X &= X' - X^0 \text{ (new minus old equilibrium quantity)} \\ \Delta Y &= Y' - Y^0 \text{ (new minus old equilibrium quantity)} \end{aligned}$$

Note, however, that the two equivalent measures of W, (i) and (ii), are more general than (iii), and that for (iii) to be an accurate measure it is crucial that in each market the initial and shifted demand functions be parallel. That is, it is necessary that the cross partial derivatives be not only equal but also constant.[3]

[3] Note, first of all, that the cross partials of the demand functions are equal only if the income effects contained in them are zero. This is a necessary condition for consumers' surplus measures (i) and (ii) to hold. Parallel shifts of the demand functions require, in addition, that the cross-partials be constant.

More generally, and with non-linear demand curves the measure will become

$$\text{(iv)} \quad \int_{X^0}^{X'} \upsilon(X, Y^0)dX + \int_{Y^0}^{Y'} \omega(X', Y)dY + x^0X^0 + y^0Y^0 - x'X' - y'Y'$$

where υ and ω are the inverses of the joint demand functions ψ and ϕ, respectively. In other words, υ and ω express price as a function of quantity rather than vice versa. The computation of these integrals will require a complete specification and estimation of the two interdependent demand functions. Since reality may not be conveniently simple, we have to face the fact that the measurement of benefits to consumers in cases requiring the use of (iv) may require considerable work.

SELECTED BIBLIOGRAPHY

Adler, Hans A. *Sector and Project Planning in Transportation*. World Bank Staff Occasional Paper No. 4. Baltimore: Johns Hopkins Press, 1967.

Beesley, M. E. and Walters, A. A. "Some Problems in the Evaluation of Urban Road Investments." *Applied Economics*, Vol. 1, No. 4 (1970).

de Weille, Jan. *Quantification of Road User Savings*. World Bank Staff Occasional Paper No. 2. Baltimore: Johns Hopkins Press, 1966.

Foster, C. D. "Surplus Criteria for Investment." *Bulletin of the Oxford University Institute of Statistics*, Vol. XXII (1960).

———. *The Transport Problem*. London: Blackie and Son, 1963.

Harral, C. G. *Preparation and Appraisal of Transport Projects*. Washington, D.C.: U.S. Department of Transportation, 1968.

Hicks, J. R. *A Revision of Demand Theory*. Oxford: Clarendon Press, 1956.

Mohring, H. and Harwitz, M. *Highway Benefits, an Analytical Framework*. Chicago: Northwestern University Press, 1962.

Samuelson, P. "Constancy of the Marginal Utility of Income." *Studies in Mathematical Economics and Econometrics*. O. Lange, F. McIntyre and T. O. Yntema (eds.), Chicago: Chicago University Press, 1942.

———. "Spatial Price Equilibrium and Linear Programming." *American Economic Review*, June, 1952.

Walters, A. A. *The Economics of Road User Charges*. World Bank Staff Occasional Paper No. 5. Baltimore: Johns Hopkins Press, 1968.

———. "The Theory and Measurement of Private and Social Costs of Highway Congestion." *Econometrica*, Vol. 29. (1961).